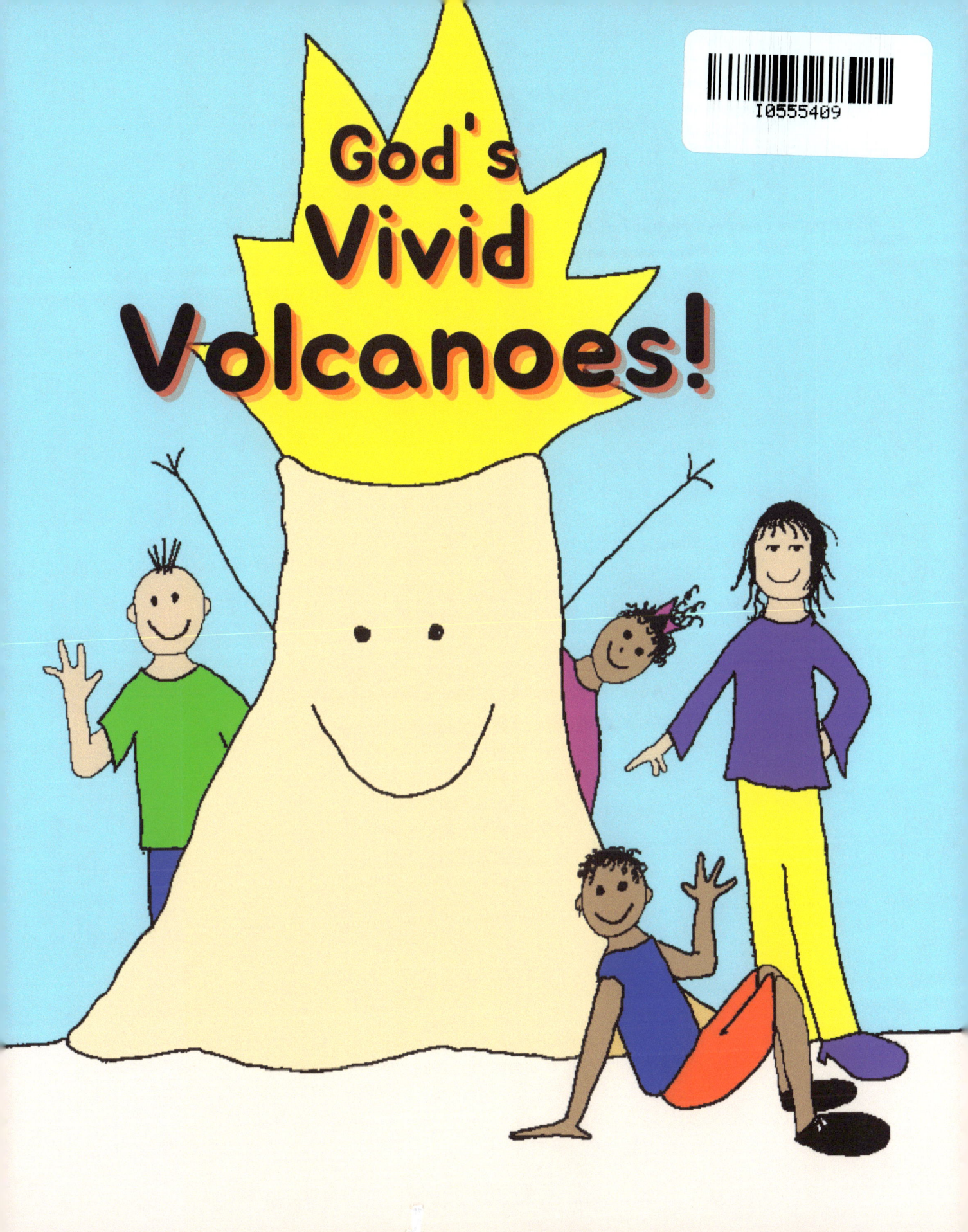

God's
Vivid
Volcanoes!

God's Vivid Volcanoes
Book 5 of the God's Cool Creation Series

Copyright © Mary Ann Winslow 2022
ISBN: 9798218123338

Author and artist - Mary Ann Winslow, PhD

Consultant - Benjamin R Winslow

Cool Creation Press
Prescott, Arizona
coolcreationpress@gmail.com

godscoolcreation.com
Instagram.com/godscoolcreation

Disclaimer: This book contains general science information intended to help the reader to better understand basic science principles. It is understood that some specific details have been omitted.

This book is dedicated to all the junior volcanologists out there! God loves you!

God created
VOLCANOES!

A volcano is a mountain
or hill with a hole,
or crater [KRAY turr],
on the top
where lava or rocks
burst out, or erupts.
There are over 1500
active volcanoes
on Earth!
But don't worry!
Most are on
the ocean floor.

Do you like volcanoes?
You can become a
volcanologist
and study these
amazing formations.
Volcano comes
from the word Vulcan,
who is the Roman
god of fire.
That's silly,
of course, because
there is only
one God!

What causes
a volcano to erupt?
Well, cracks in the
Earth made of large
slabs of rock,
called tectonic
(Tek TAHN ik)
plates, sometimes
move and cause an
eruption.
OR...

Pressure!
Deep inside is a
very hot liquid called
magma
(outside, it's called lava!).
Pressure can
build up
from trapped gasses
and it has to go
somewhere,
so it bursts through
the opening on the
mountain.

Lava
can also flow
easily down
like a river
and not shoot out at all!
If the inside is
more watery,
then it flows like
a river.
If the inside is
more sticky,
with more rocks,
it bursts!

Magma is lighter
so it rises to the top.
Sometimes other stuff, like
boulders, cinders, and ash,
fly out, too.
Gas bubbles can cause
rocks, called pumice,
to have lots of holes.
Pumice can actually float
on water!
These rocks can split
up into tiny pieces,
called cinders.

God created mainly
four types of volcanoes.
A shield volcano gently,
quietly flows
through the tube,
called a vent, inside.
It flows in all directions
and it flows the farthest!
Composite, or
stratovolcanoes,
are the biggest
because over time
what spits out – lava, rocks,
and ash – builds up around it.
They're also the noisiest!

A cinder cone
volcano
tells you what it spits
out - tiny pieces
of rock
called cinders!
Lava domes,
or volcanic
domes, are mounds
too thick to travel
very far.
They just create
mounds that
"grow" around it.

When a composite
volcano collapses,
a caldera [call DAYR uh]
forms, which
is just a big opening.
Lakes can form
in them, like
Crater Lake in Oregon.
Composite volcanoes
make up the
Ring of Fire.
They form a ring or chain
around an area in the
Pacific Ocean
that looks like
they're all connected.

Volcanoes
break down
over time
due to erosion
[ee ROW zhun]
and become
a neck, or plug.
Ship Rock
on the
Navajo Reservation
in New Mexico
is a good
example of one.

Flat sheets of rock
on the sides of mountains
are called sills.
They form when magma seeps
between rock layers.
Pointy sheets of rock
are called dikes.
They form when magma
finds its way
through a crack of rock.
Dikes burst out
in different directions
when Ship Rock
erupted.
Picacho Peak
in Arizona
is also considered a dike.

God created volcanoes
for a purpose!
They bring rich soil
that allows
vegetables and fruits
to grow.
Their material is also used
in roads,
roofs, cement
and concrete,
and in soaps
and cleaners!

Steam from
volcanic activity
helps to warm
really cold climates
like Iceland.
Volcanoes
add water
to our air,
they make islands,
they produce water
in our oceans,
and more!

God made
interesting
and
helpful
volcanoes
all over the world
just
for us!

Thank you, LORD!

www.ingramcontent.com/pod-product-compliance
Lightning Source LLC
Chambersburg PA
CBHW041446120626
46547CB00002B/361